PLC Course

Beginning PLC Basics

Introduction to PLC

PLC COURSE

-

- ISBN-13:
- 978-1500121327

-

- ISBN-10:
- 1500121320
-
-

- Brian Daniel Starr is the owners of this work with all rights and responsibilities related to this work.

-

PLC 101

Beginning PLC Basics

Introduction to PLC

PLC

- PLC stands for Programmable Logic Controller
- Basic Course will focus on beginning of what a novice needs to know
- Key word Programmable
- Beginners use Ladder Logic based on Electrician Drawings

PLC I/O

- Keyword is Controller
- PLC uses logic and state of Inputs to control Outputs
- Logic written with ladder logic
- PLC scans program many times over and over
- Inputs can be analog or digital

Digital Versus Analog

- Digital means on or off

- Usually Digital is signal voltage

- Inputs are on if voltage on terminal

- Outputs turn on Voltage sourced from PLC or close switch for Voltage sourced from Power supply run thur switch on PLC

- Analog means a value usually from 4 to 20 ma or from 0 to 10 volts

Digital Versus Analog

- Digital can be expressed as a binary number good for computers

- 1 is on and 0 is off

- Analog is expressed as a span from 0 to 100 percent or however it is scaled from 4 ma to 20 ma etc.

- Analog has an accuracy based on how many steps between min and max, so if there are 32678 steps between 4 and 20 ma then 16 ma divided by 32678 leaves a resolution of on half of one micro amp per step. (approximately)

Beginners Basic Demonstration

- Input one is a wire from a switch

- Output one goes to a light bulb

- Now you could wire the switch to the light bulb, however the logic would be if switch on then light bulb on

- So the PLC determines that SW1 is for LI1 if SW1 is the Switch and LI1 is the Light

Here it is in Ladder

- SW1   LI1
- --| |----------------------------------()-----

Wire another Switch

- Now add SW2

- If SW2 or SW1 then cut on LI1

- SW1   LI1

- -| |----|--------------------()----

- -| |----|

Again with SW2

- Now we want the light on only when both switches are on

-    SW1   SW2   Ll1

- ---||------------||------------()

Notice we can make it happen with no rewiring, just reprogramming

- Here is input one and input two wired to switches

- Here is output one wired to light

- Programmer can make light come on if either switch is on

- Or Programmer can make light come on only if both switches are on

Concept

- Programmable in that switches are wired to PLC, and Light wired to PLC, both the same

- Can program for either or both

- Imagine if there are 100 switches and 100 lights

- Now Imagine a field of 100 led lights in a ten by ten pattern

- Programming letters possible etc.......

PLC programs

- PLC programs are written in ladder segments
- One main program is usually used to call all the others
- The program executes, then starts over again, usually many thousand of times each second
- One scan is one process.
- PLC usually look at input table, then process the logic, then look at output table

PLC program example

- Station One Pick and Place

- Imagine a cylinder on a clamp

- Imagine a clamp on an Up down cylinder

- Imagine an Up down cylinder on a traverse cyclinder going in and out

- Now imagine a part bin and a table that is round that rotates

- Idea, pick up part from part bin and put on table

Here is programming Algorithm

- Outputs are clamp close, cylinder up, cylinder down, traverse arm in, traverse arm out

- Steps. 1. go down on part 2. clamp on part

3. Go up on part, 4. go into table (traverse), 5 go down on table. 6 unclamp part 7. go back up. 8 go back to part bin. 9 wait for next cycle

4. Simple pick and place

PLC ladder rung routine

- Command to Turn on Clamp then wait 1 second
- Command to Go Up wait 1 second
- Command to go in then wait 1 second
- Command to go down on table wait 1 second
- Command to release wait one second
- Command to go up wait one second
- Command to go over part bin wait one second

Ways to improve performance

- Instead of command then a timer to wait, command and have a sensor sense that clamp in, or clamp up, or clamp over table etc...

- Notice now there are output, clamp, up, over, down, and unclamp, go back.

- Notice a sensor needed for clamped, unclamped, up, down, over table, over part bin

Now Imagine table turns

- If the table is simple, four stations, one is add part one, two is add part two, three is weld parts together, four is unload table

- Now add PLC output turn table

- When table turned have a sensor saying everything in place

- Station one adds part one, Station two adds part two, stations very similar can use same code.

Table Turns continued

- Station three is welder that drops on part, welds parts together. And then goes up out of the way

- Station four simply unloads part. Similar to pick and place except drops part in chute. Picks up part from table

- PLC has then five stations or operations

Five Stations

- 1. add part one (runs stations 1 thru four together)

- 2. Add part two

- 3. Weld together

- 4 Unload table

- 5. Rotate when operations 1,2,3, and 4 are finished and then start over again when all are ready

Hazardous or Toxic

- Machine can be monitored with camera

- Explosives or Toxic chemicals can be handled with automation that would put a human at risk

- PLC can program this with controls for dismantling explosives or toxic chemicals or other demilitarization

Cost Effective

- A small PLC can cost about 100 to 500 dollars

- One timer for a process costs 100 dollars

- One Switch can cost about 100 dollars

- Wiring a machine of 10 switches, 10 timers and 10 motors can cost a lot and cannot be modified

- Total cost of 10 timers is 1000 dollars

- PLC cost is 500 with unlimited timers

- Cost the same for switches, even if wired to timers and motors or wired to PLC

PLC and RTU

- PLC actually trademark name. RTU is used as well.

- Either case a small PLC can be the same as an RTU

- The PLC or RTU can have unlimited inputs and outputs

- PLC and RTU can be used on many types of projects

Simple Projects for PLC or RTU

- Keeping tanks full for City water system or industrial process etc.

- Input analog for level of tank

- Programmable selection of level when to turn on pump to refill tank

- One PLC can be for an entire Tank Farm at the refinery or for an entire City water system

Simple Conveyor

- Inputs Photo eye on conveyor section for sensing if box present

- Lets Say conveyor has ten sections

- Output is motor on off for ten sections of conveyor (10 motors)

- If Box on section don't add another box, if no box then add box by turning on motor on previous conveyor etc......

- Notice all control is completely digital

PLC interface with HMI

- If one switch is 100 dollars, 25 motor conveyor is 2500 dollars in switches alone.

- Cheap HMI and PLC for 500 dollars and HMI for unlimited switches for 500 dollars can turn on and off all motors. photo eyes, manual and auto control, report errors etc.... With large hardware cost saving and is easily modified

Why PLC and Automation?

- Many processes require and inspection or a judgement based on putting an operator in a plant to make a decision and press a button or two.

- PLC and automation can make the decision and press the button for the operator, freeing the operator to do something else

Cost of Operator

- A trained operator in a plant costs at minimum wage about 40,000 dollars a year

- Replace one operator with automation on a machine that lasts ten years requires engineering investment less that 400,000 dollars to be cost effective

- If investment only 100,000 dollars then company saves 300,000 dollars over ten years

Cost of Automation Effective

- If an engineer can replace 10 minimum wage operators at a cost of 1 million dollars and the company is planning to run for ten years

- Cost of 10 operators @ 40,000 per year over ten years is 4 million

- Cost of Engineering and hardware 1 million

- Savings 3 million dollars

Add Vision

- Visual inspection on high speed lines replaces operators and inspectors

- Visual inspection interfaces well with PLC

- Small Vision systems cost effective

- Camera's can see things and inspect things that human operators cannot

Add Robots

- Robots are easily programmable

- Robots interface well with PLC

- Robots replace humans on two motors or on shipping lines well

- Robots are easily programmable

- Robots are servo's working together on axis,

- Servos' are easily controlled by PLC

PLC ties it all together

- The PLC is the central focus of the computational ability of the system

- Five robots, data logging, HMI, and one PLC to tie it all together is an example system

- PLC if not the central controller certainly an important part

PLC now very complex

- PLC was small in the beginning

- PLC now expandable to many thousands of inputs and outputs and specialty cards

- PLC still robust, does not fault, can run years without failure

- PLC primarily programmed in machine code

- PLC interface with anything

Single PLC for machine Control

- A simple machine that makes widgets could have a few operations, great programming for PLC programmer

- Machine will output widgets for years in a plant if given parts to assemble

- Great task for small PLC

Specialty Cards for PLC

- Almost every device has a specialty card so it can interface with the PLC

- The PLC is modular, it has input cards, output cards, and specialty cards.

- Inputs must be signal dc voltage or ac voltage usually 24 v dc or 110, 220 or 460 ac

- Outputs could be signal or just a switch closure called dry contact, can be used for any voltage

Specialty Cards

- Specialty Cards are communication cards, device cards such as interface for vision or hart meter etc.

- Special counter cards for high speed counting such as used with encoders

- Cards to be used with servos for motion

- Special diagnostics can be built in

PLC logic

- Time for switches and actuators to move things etc must be a program consideration

- If you turn on an output to extend a cylinder, put a sensor at the end of the cylinder to turn off the output.

- If no sensor on cyclinder you can predict it will happen in a time period and turn off output after a time has elapsed.0

PLC Large control versus small Control

- Large PLC good for large process like plant assembly line

- Small PLC good for single process like single machine or one station of assembly line

- Small PLC good for single machine control

- One large PLC can be interfaced with many small PLC

- PLC's can network like computers (PC's)

Conclusion

- Introduction of PLC shows how automation is cost effective

- PLC can be used for small or large scale automation

- PLC programming a skill that is easily marketable

- PLC and automation part of Manufacturing and will never go away

PLC 102

SEQUENCER

Sequencer Definition

- Sequences thru steps that both have an action when active and have a transistion when complete

- Each step will trigger the action or event

- The Action or event will end with for Example a sensor or a timer finishing

PICK and PLACE

- The pick and place mechanism is a device to show how a sequencer works

- The mechanism will consist of three cylinders actuated by air.

- 1   Close/open clamp

- 2   Travel Up/Down

- 3   Travel In/Out

Transistions

- Clamps are difficult to put sensors on so we will use a timer to say the clamp has actuated

- The Up Down Cylinder will have a sensor that says the Cylinder is Up and a sensor that says the Cylinder is Down

- The In Out Cylinder will have a sensor that says the Cylinder is out or the Cylinder is in

Concept for Pick Place

- The mechanism will go down and PICK up the part
- The mechanism will go up
- The mechanism will go out over the place point
- The mechanism will go down
- The mcahnism will open and PLACE the part on the Place point, then will return

Return Path

- The Return Path is to go up after leaving the part at the PLACE point

- After up the mechanism goes In

- After it is in it waits above the PICK point until it is time or get another part

- Or in some cases it saves time to go down and PICK the part, and then Go Up and wait

- In this example we will wait with an empty clamp

Making Variables or Tags

- We will need a few Registers for the Inputs, Outputs, Timers, and Variables or Tags

- Up will be Input I:1.0

- Down will be Input I:1.1

- In will be Input I:1.2

- Out will be Input I:1.3

- The Register will be an integer register and will be called Step

The Outputs will be single Acting

- There are three Cylinders.   These will have one output for ON and extend and if Off will retract

- The Outputs will be O:1.0 for clamp (grip is on)

- O:1.1 for Down, and off for Up

- O:1.2 for Out and off for In

- Thus if all off the Grip is open Up and In.

Note on Tags

- Older PLC did not have tags that are created and used registers directly

- For this example we will use RS Logix 5000 for the programming so we can use any names for the tags

- We will use the Tags that are the Input and Output Tag Registers so the training will work as well with the Older SLC 500 and PLC 5

List of Tags Used

- I:1.0, I:1.1, I:1.2, I:1.3, O:1.0, O:1.1, O:1.2, Timer1, Timer2, Step

- Coorresponding to these tags is the following list

- Up, Down, In, Out, Clamp, Down, Out

Steps Defined

- Step 0   Wait
- Step 1   Go Down
- Step 2   Clamp
- Step 3   Go Up
- Step 4   Go Out
- Step 5   Go Down
- Step 6   Release Clamp
- Step 7   Go Up
- Step 8   Go In

Step 0

```
                              ***
                             *****
                    Wait for Cycle On Switch
                      Progress to Step 1
                             *****
                              ***
                               *

             I:1.4  Cycle On
    EQU      Local:1:I.Data.4                          MOV
   Equal                                              Move
   Source A   STEP   ] [                              Source    1   Dest   STEP
               0                                                              0
   Source B    0
```

Step 1

```
                           +
                          +++
                         +++++
        Send Clamp Down and Wait For Sensor
                Progress to Step 2
                         +++++
                          +++
                           +
```

```
                      I:1.0 Cylinder One
                         is Down
                      Local:1:I.Data.0
  ┌─────EQU─────┐          ] [                      ┌─────MOV─────┐
  │ Equal       │                                   │ Move        │
──┤ Source A  STEP│────────────────────────────────┤ Source    2 │── Dest   STEP │──
  │           0 ←│                                   │          0 ←│
  │ Source B    1│                                   └─────────────┘
  └─────────────┘
```

Step 2

Close Clamp. Wait for Timer that assumes Clamp is Closed

Progress to Step 3

```
                                                                                          MOV
         EQU                  TIMER1.DN                                                    Move
         Equal                                                                             Source    3     Dest    STEP
2        Source A    STEP                                                                                             0
                        0                                                                                             ←
         Source B     2
```

Step 3

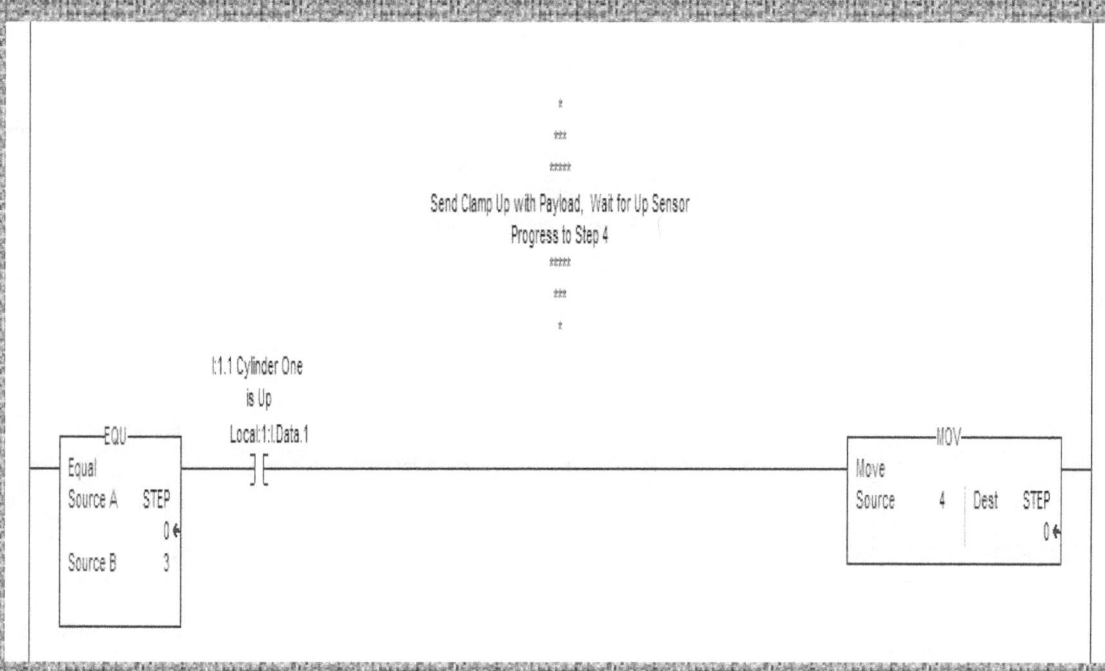

```
                                    *
                                   ***
                                  *****
              Send Clamp Up with Payload,  Wait for Up Sensor
                          Progress to Step 4
                                  *****
                                   ***
                                    *

                          I:1.1 Cylinder One
                               is Up
                          Local:1:I.Data.1
     ┌──EQU──────────┐         ┤ ├                        ┌──MOV──────────────────┐
     │Equal          │                                    │Move                   │
     │Source A   STEP │                                    │Source      4  Dest STEP│
     │            0 ← │                                    │                     0 ←│
     │Source B      3 │                                    └───────────────────────┘
     └───────────────┘
```

Step 4

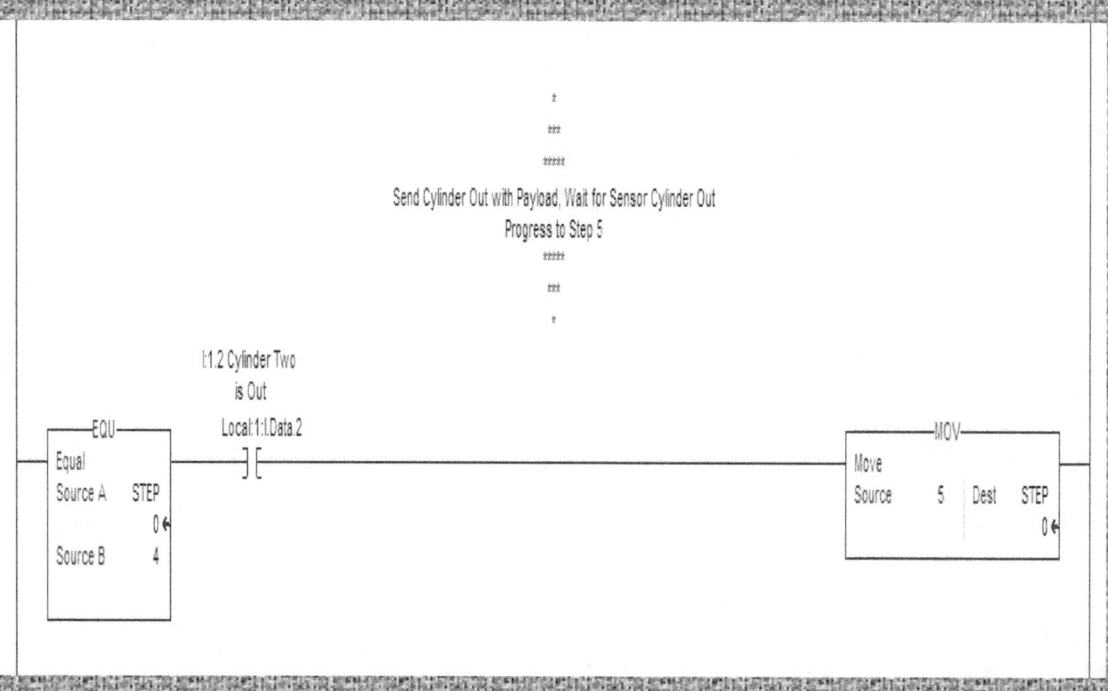

```
                                    *
                                   ***
                                  *****
         Send Cylinder Out with Payload, Wait for Sensor Cylinder Out
                          Progress to Step 5
                                  *****
                                   ***
                                    *

                    I:1.2 Cylinder Two
                         is Out
                    Local:1:I.Data.2
    ┌──EQU─────────┐                                      ┌──MOV─────────────┐
    │ Equal        │      ─┤ ├─                           │ Move             │
    │ Source A  STEP│                                     │ Source      5    │
    │            0 ←│                                     │ Dest      STEP   │
    │ Source B    4 │                                     │             0 ←  │
    └──────────────┘                                      └──────────────────┘
```

Step 5

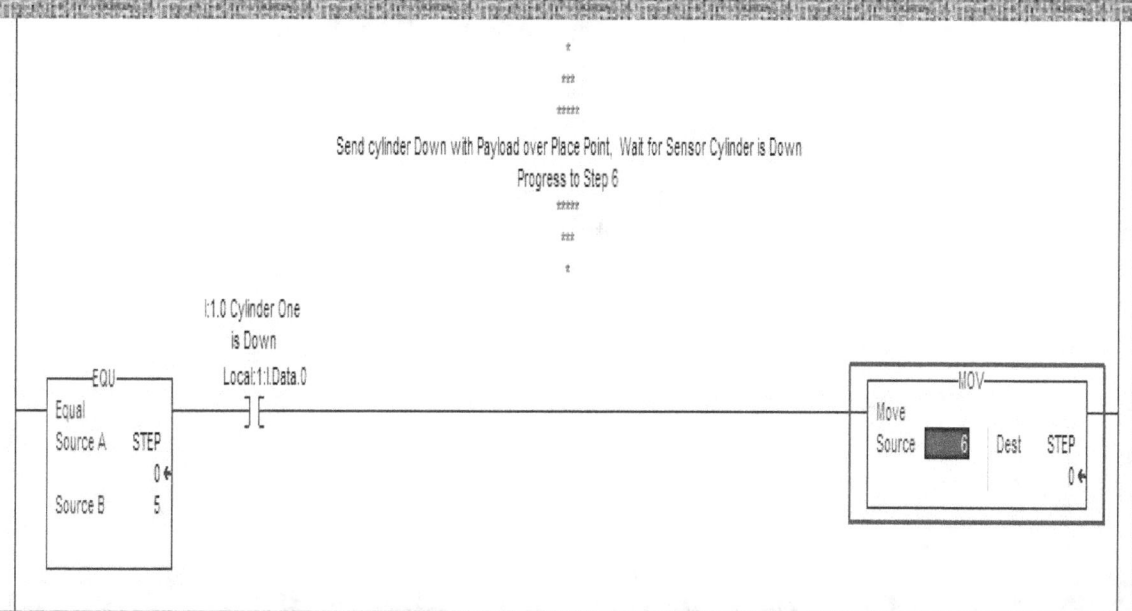

Send cylinder Down with Payload over Place Point, Wait for Sensor Cylinder is Down
Progress to Step 6

I:1.0 Cylinder One
is Down
Local:1:I.Data.0

EQU
Equal
Source A   STEP
0
Source B   5

MOV
Move
Source   6   Dest   STEP
0

Step 6

Unclamp and release Payload on Place Point, Wait for Timer to expire saying Clamp Open
Progress to Step 7

```
      EQU                    TIMER2.DN                                          MOV
   Equal                                                                     Move
   Source A   STEP         ] [                                               Source      7    Dest   STEP
                0 ←                                                                 ←                      0 ←
   Source B     6
```

Step 7

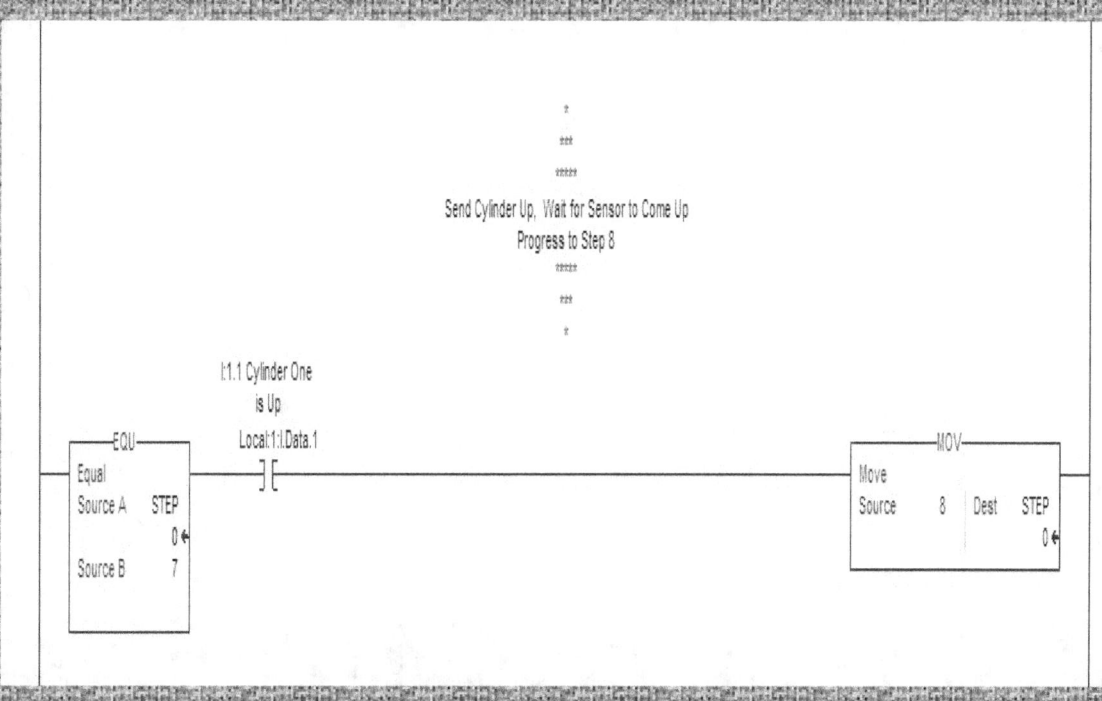

```
                              *
                             ***
                            *****
        Send Cylinder Up,  Wait for Sensor to Come Up
                   Progress to Step 8
                            *****
                             ***
                              *

                  I:1.1 Cylinder One
                       is Up
                  Local:1:I.Data.1
    --EQU--                                              --MOV--
    Equal             ] [                                Move
    Source A    STEP                                     Source    8    Dest    STEP
                  0←                                                               0←
    Source B      7
```

Step 8

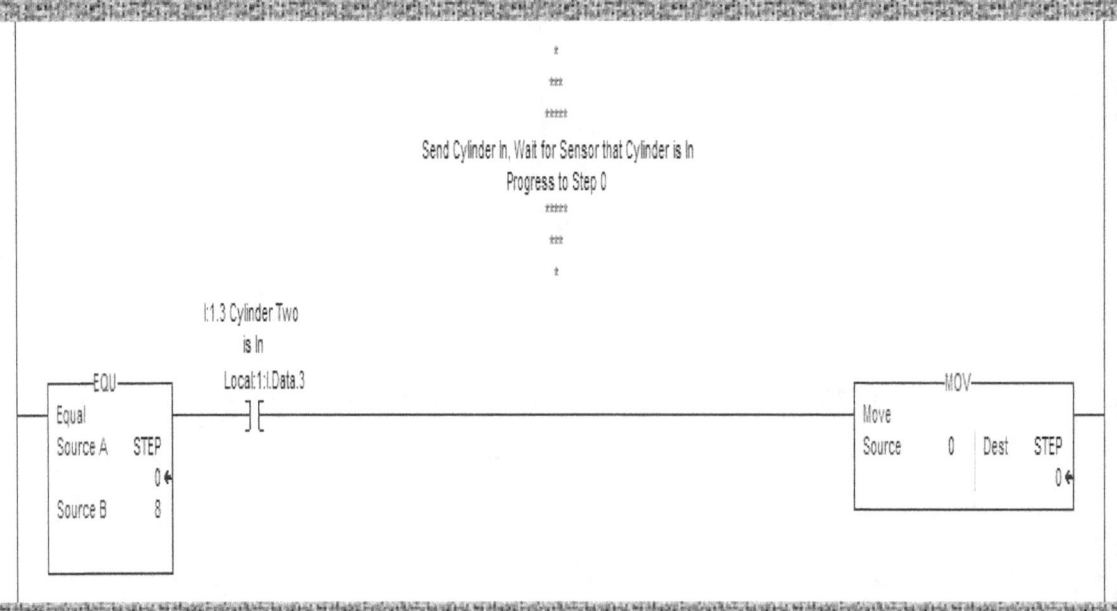

Send Cylinder In, Wait for Sensor that Cylinder is In
Progress to Step 0

Actions

- For Each Step something must be actuated or a timer started

- In PLC programming it is customary that only one instance of an output is ever entered

- In other words, you cannot have the same output in two places

- Thus if an output such as Down is in more than one place there are two conditions to make it work.

Actions Continued

- Also if the action is to be out for more than one step the output must have more than one step to actuate the output

- Because the output coil is only entered in one place the steps are entered in parallel on the rung

Clamp

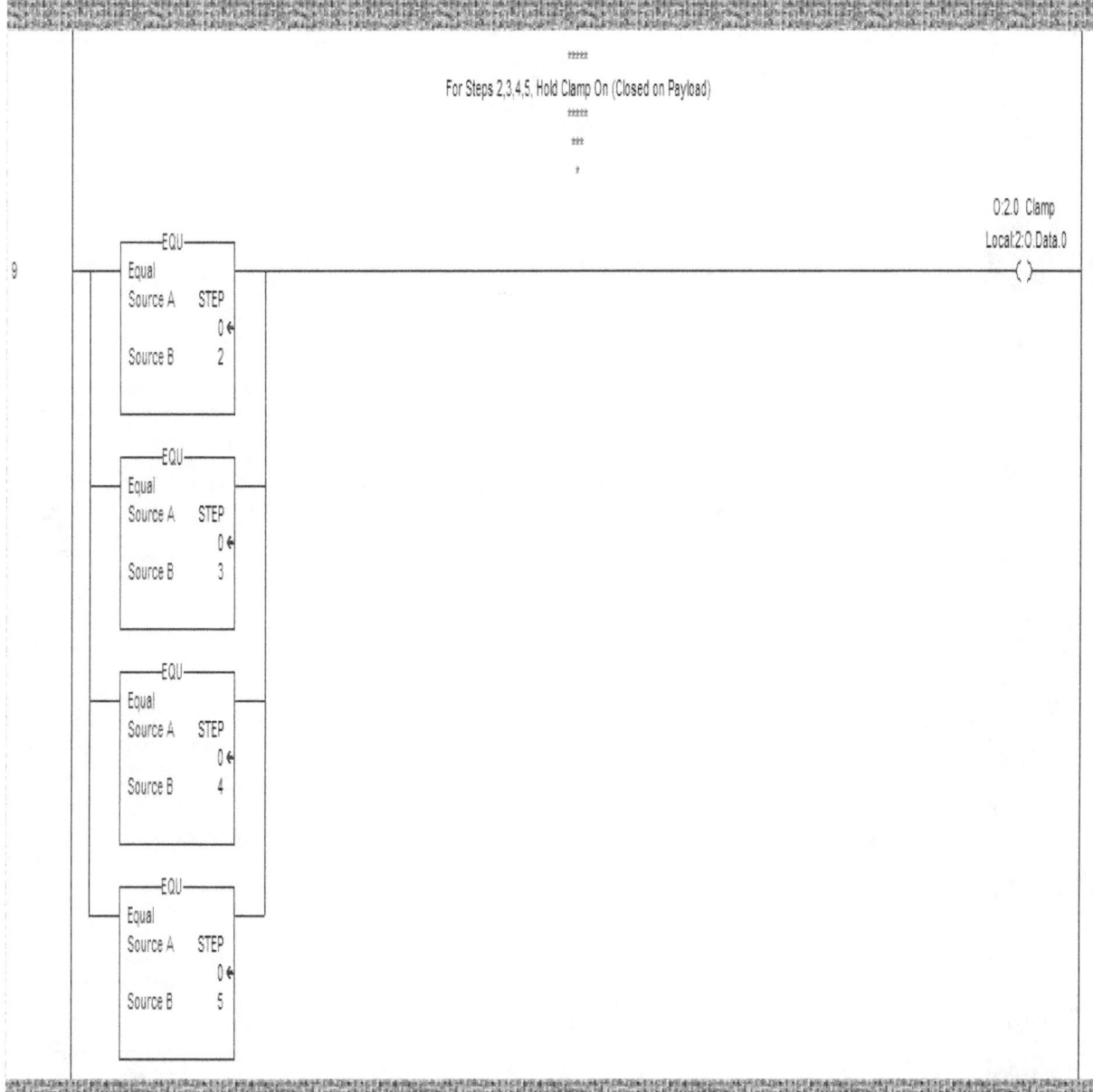

For Steps 2,3,4,5, Hold Clamp On (Closed on Payload)

O:2.0 Clamp
Local:2:O.Data.0

9

EQU
Equal
Source A   STEP
0
Source B   2

EQU
Equal
Source A   STEP
0
Source B   3

EQU
Equal
Source A   STEP
0
Source B   4

EQU
Equal
Source A   STEP
0
Source B   5

Down

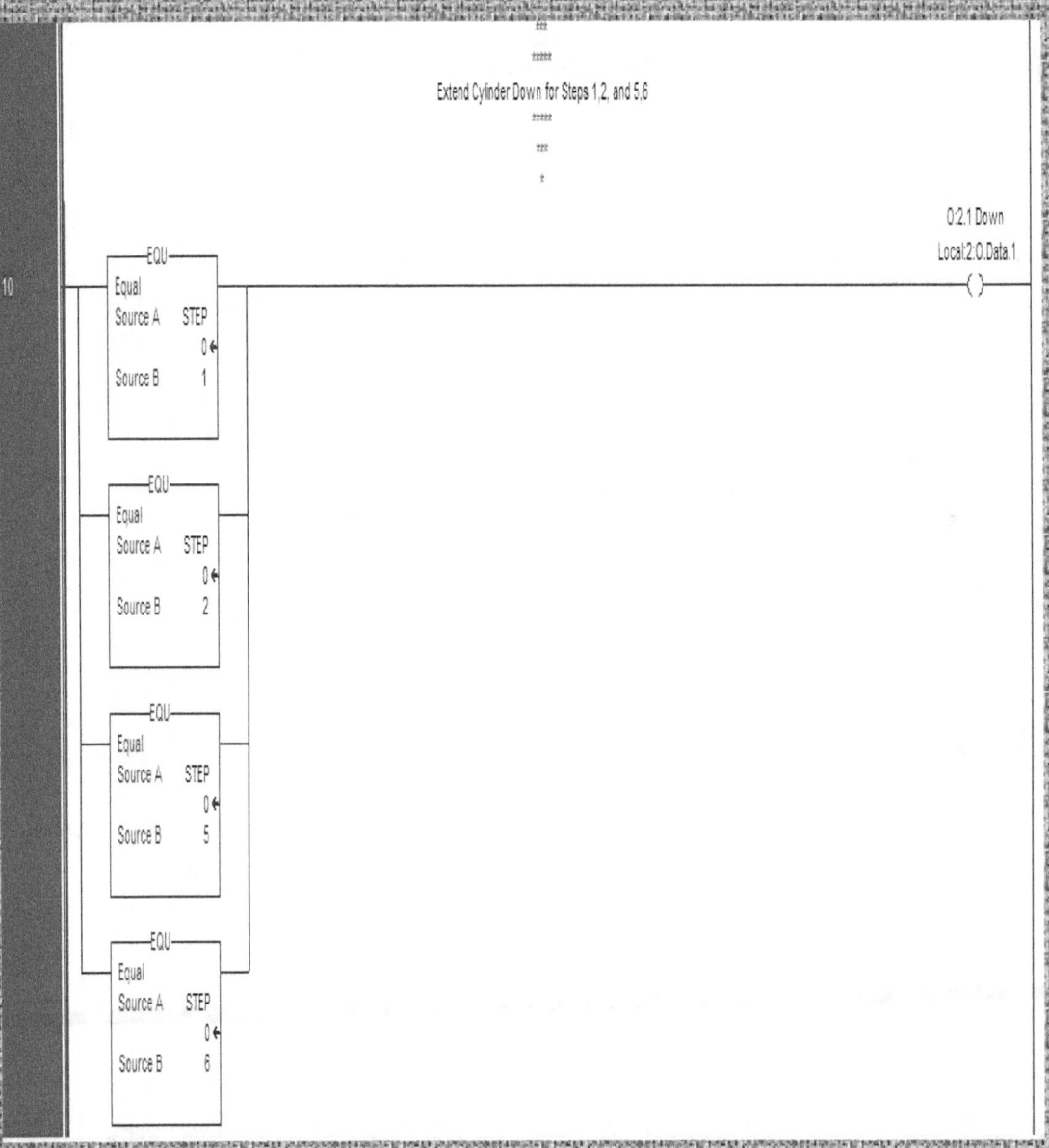

Out

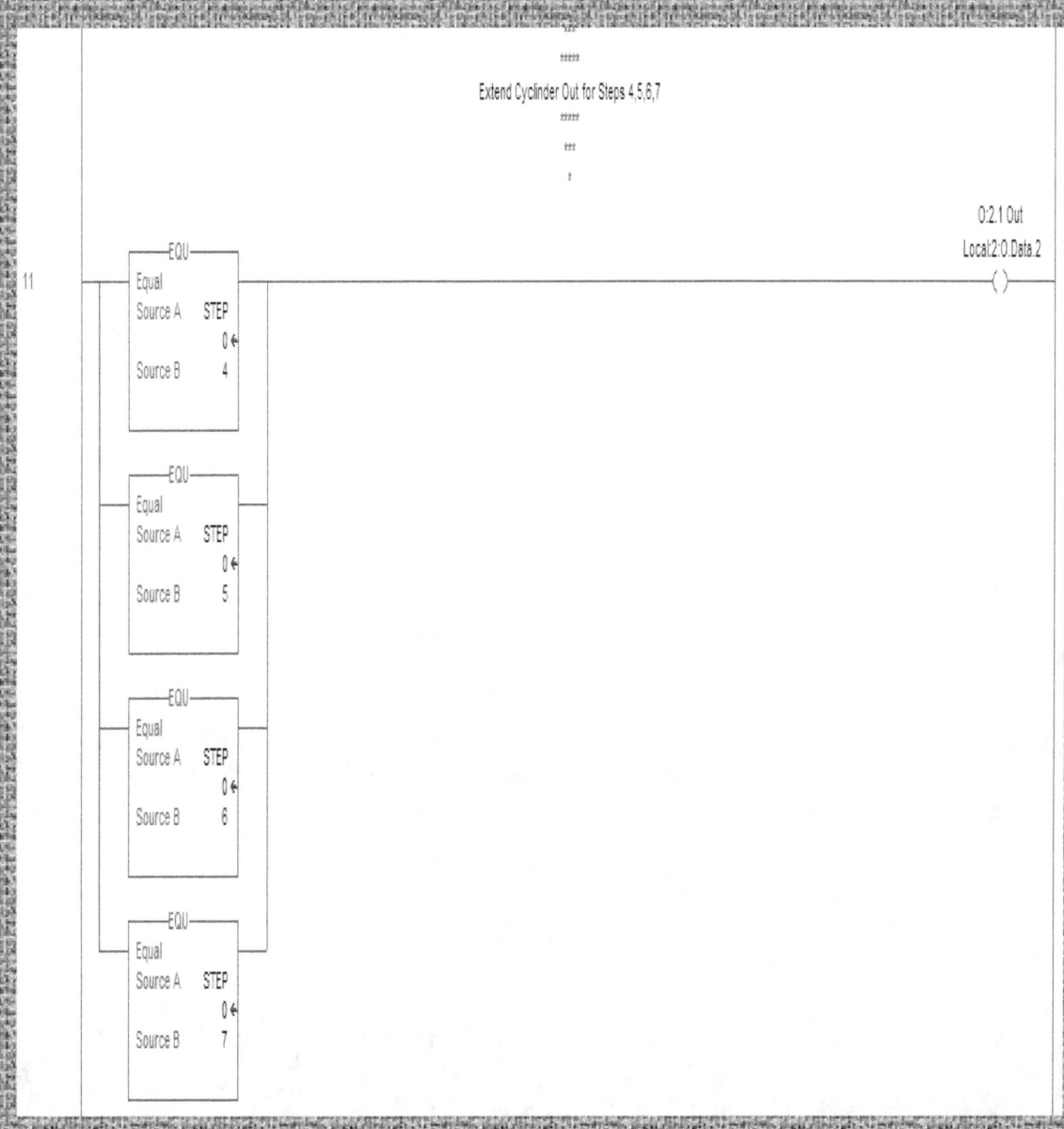

Extend Cyclinder Out for Steps 4,5,6,7

O:2.1 Out
Local:2:O.Data.2

11

EQU
Equal
Source A   STEP
            0
Source B   4

EQU
Equal
Source A   STEP
            0
Source B   5

EQU
Equal
Source A   STEP
            0
Source B   6

EQU
Equal
Source A   STEP
            0
Source B   7

Timer One

```
                    *
                   ***
                  *****
   Timer One is on for one half second (500 miliseconds) for Clamp to close
                  *****
                   ***
                    *

        ┌────EQU────┐                                    ┌────TON────┐
        │ Equal     │                                    │ Timer On Delay    ─(EN)─
 12 ────┤ Source A    STEP ├──────────────────────────────┤ Timer    TIMER1   ─(DN)─
        │                0 ←                               │ Preset      500 ←
        │ Source B      2  │                               │ Accum         0 ←
        └───────────┘                                    └───────────┘
```

Timer Two

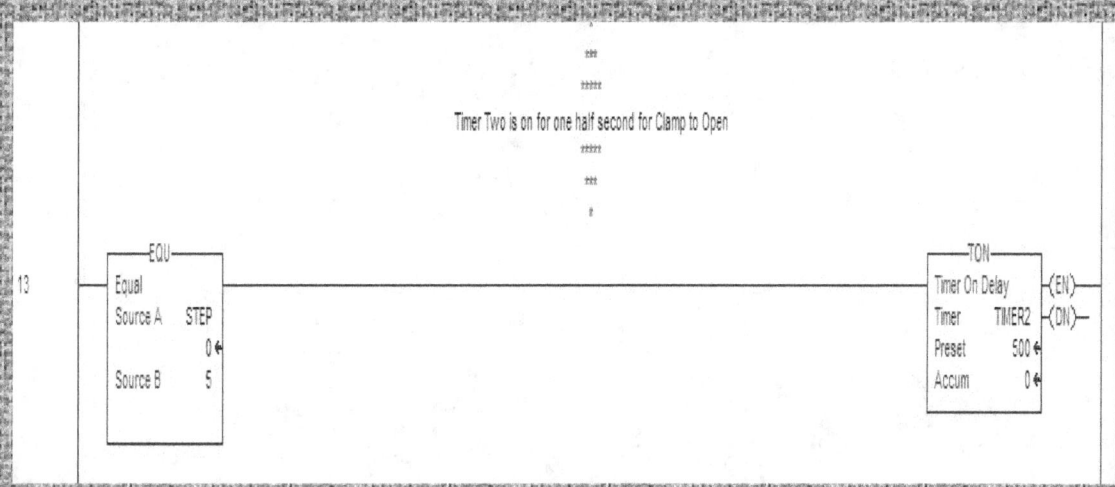

Timer Two is on for one half second for Clamp to Open

	EQU			TON	
13	Equal			Timer On Delay	⟨EN⟩
	Source A	STEP		Timer	TIMER2 ⟨DN⟩
		0 ←		Preset	500 ←
	Source B	5		Accum	0 ←

Notice All Steps Not Used

- Because the cyclinders return when they are not on, the steps 3 and 7 are go Up or turn off the Down, the steps 8 are go In or turn off the Out and the steps 6 are unclamp or turn off the clamp.

Contact for Cycle On

- This contact is the contact for step 0 to go to step 1

- If this contact copied to every step then the process can be stopped at any step by opening the Cycle On contact

Typical Pick and Place

- The Pick and Place is a basic building block of machine control

- A Machine will usually have a few pick and place and then an operation to put the parts together

- A Pick and Place can be used for picking up parts and putting on a machine as well as picking up parts from the machine and putting them in output bin off machine.

Many Station Machine

- A machine will have a sequencer for each station on a machine

- A sequencer will be triggered as well for turning the table on a machine

- Each pick and place fills a station; then the table rotates and it does it again

- A 25 station machine will cycle 25 times before a complete part is made, but then everytime it cycles it makes a new part.

PLC 103

Simple Fill Tank

Application

PUMP and TANK

- For this simple application we will discuss a Tank with a Pump

- The Tank will have an analog level Control

- The Pump will be discrete control On and Off

RS Logix 5000

- There is a RSLogix 5000 program for this routine

- The rungs will be a compare and set bit if low level

- Next, there will be a compare and set if full level

- There will be an error section

- The main rung will demonstrate a latch

Tags

- Tags will be memory Tags

- Inputs will be one analog level input that will be scaled to 0 to 100

- There will be a Low Level On

- There will be a High Level reached

- For alarms there will be LowLow, Low, High, and HighHigh Alarm

- The output will be O:2.0 Pump On

Scaling

- The scaling is done by the program, although in RSLogix 5000 could be done on the properties screen. Older PLCs did not have the capability of scaling.

```
                                    *
                                   ***
                                  *****
              Compute the Scaled Level for the transmitter
    In this example with RS Logix 5000 this could be done with a properties screen for the Input Card
 It was done in this manner in order to demonstrate how it would need to be done on an analog input card that is from 0 to 32768
                                  *****
                                   ***
                                    *
                                                                        ┌─────CPT──────────────────────┐
                                                                        │ Compute                       │
  0 ─────────────────────────────────────────────────────────────────── │ Expression  Local:1:I.Ch0Data * │ Dest    Level │
                                                                        │             (100 / 32768)     │         0 ← │
                                                                        └───────────────────────────────┘
```

Determine Low Level

- Low Level is reached when the level drops below 25 percent

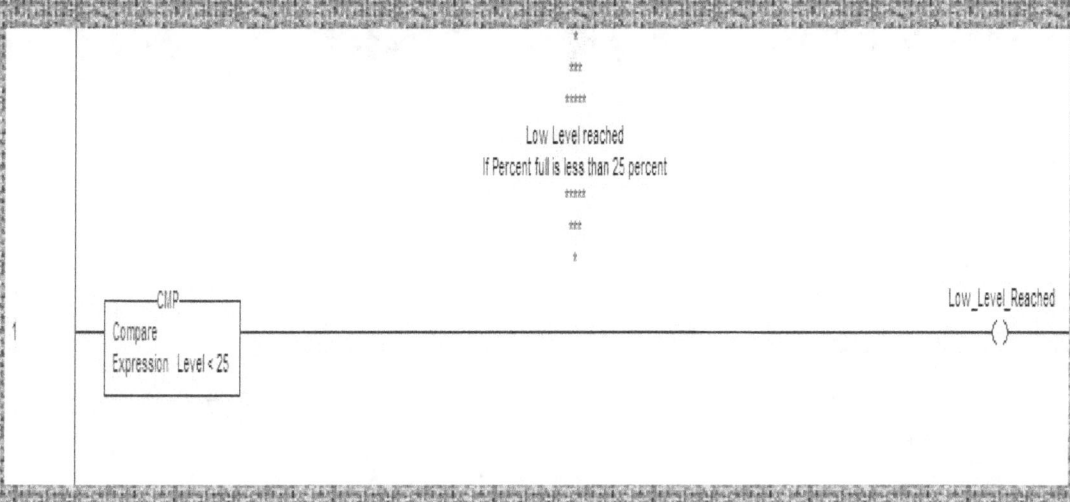

High Level Reached

- The High Level is a compare that is made when the level is over 75 percent

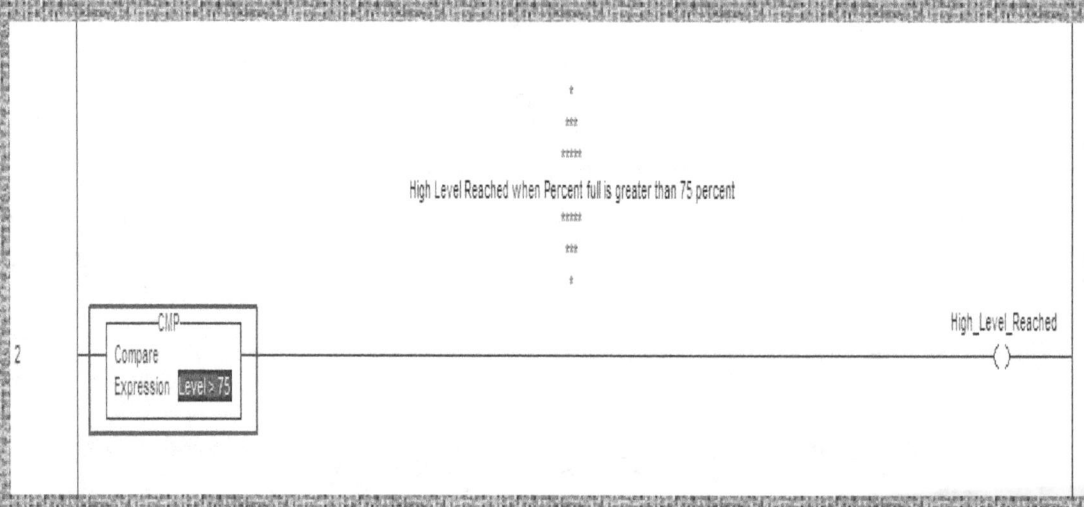

Latching Rung

- Condition makes it come on, the branch of the output seals the latch, and the examine off for High level unlatches the rung

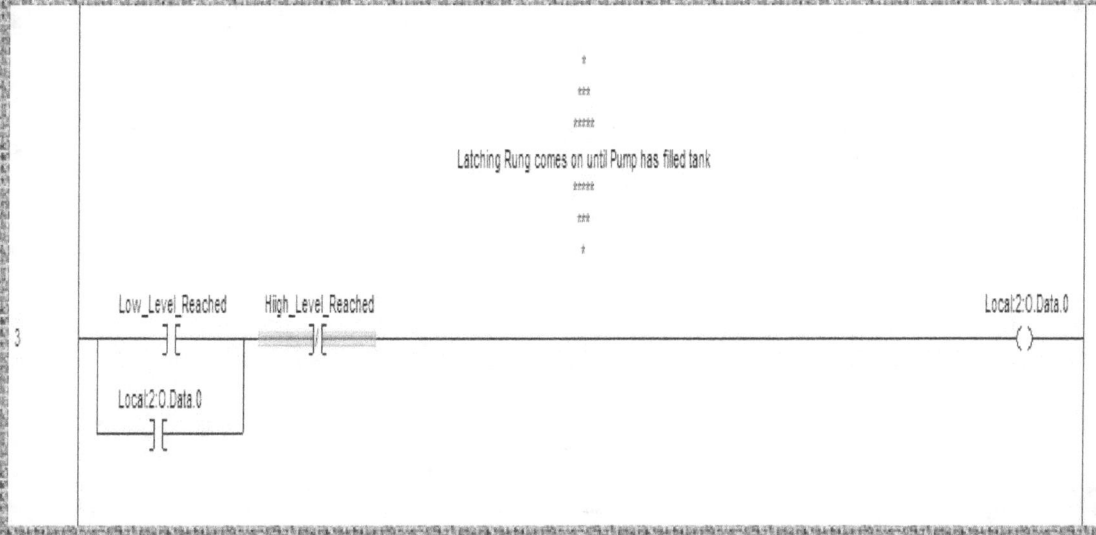

Alarms

- Alarms can be picked up by HMI

Alarms

- The Low Level and High Level are not critical, but if the pump is operational and the level control works, these alarms cannot ever occur unless there is a problem

- High High level reached may mean an overflow is pending

- Low Low level may mean the pump could cavitate if not able to pump air and damage to equipment might happen.

Jump to Subroutine Call

- This Jump is in the Main Routine so that the subroutine is scanned.

0		JSR Jump To Subroutine Routine Name PumpFill
(End)